3rd International Conference on Microbiome Engineering (ICME 20)

Online
7-9 December 2020

ISBN: 978-1-7138-2773-3

3rd International Conference on Microbiome Engineering (ICME 20)

Online
7-9 December 2020

AIChE
The Global Home of Chemical Engineers

Printed from e-media with permission by:

Curran Associates, Inc.
57 Morehouse Lane
Red Hook, NY 12571

Some format issues inherent in the e-media version may also appear in this print version.

Printed with permission by Curran Associates, Inc. (2021)

For permission requests, please contact AIChE
at the address below.

AIChE
120 Wall Street, FL 23
New York, NY 10005-4020

Phone: (800) 242-4363
Fax: (203) 775-5177

www.aiche.org

Additional copies of this publication are available from:

Curran Associates, Inc.
57 Morehouse Lane
Red Hook, NY 12571 USA
Phone: 845-758-0400
Fax: 845-758-2633
Email: curran@proceedings.com
Web: www.proceedings.com

TABLE OF CONTENTS

MICROBIOME THERAPEUTICS: ADDITIVE APPROACHES

MICROBIOME THERAPEUTICS: MODULATORY APPROACHES

MICROBIOME THERAPEUTICS: SUBTRACTIVE APPROACHES

Design of a Novel Poly(dimethylsiloxane)-Based Microbial Nanoculture System

Conference: Microbiome Engineering

Year: 2020

Proceeding: 3rd International Conference on Microbiome Engineering (ICME 20)

Group: General Submissions

Session: Microbiome Therapeutics: Additive approaches

Authors:

Usman, H., University of Illinois at Chicago
Davidson, S., University of Pittsburgh
Bah, A., The Ohio State University
Beckman, E. J., University of Pittsburgh
Niepa, T. H. R., University of Pittsburgh

Organosilanes contain hydrocarbon-like backbones, allowing them to react with silicone-based agents in the presence of a catalyst and polymerize into membranes with tunabletransport and mechanical properties. Owing to their high hydrophobicity, Poly(dimethylsiloxane) (PDMS) membranes, and more particularly, SylgardÂ® 184, have been used for applications including drug delivery, gas separation, and microfluidics fabrication. However, the undefined composition of the material and its ability to leach out uncured oligomers make its functionalization and usage challengingfor many biological applications. This article presents the design of a novel culture system generated using PDMS-based membranes to study microbial dynamics. The microbial culture system that is referred to as â€œnanocultureâ€☐ serves to encapsulate and grow microbes in semipermeable membranes. The mechanical properties of the membranes are investigated through osmotic annealing to determine the ability of the nanocultures to withstand high shear stress similar to environmental conditions while maintaining transport properties essential to microbial communication and growth. The present study lays the foundation for a novel microbial culture system that would enable the cultivation of microorganisms in environments other than laboratory conditions.

Sweet and Sour; Microbiota-Directed Fibre Activates Both Targeted and Secondary Metabolic Shifts in the Distal Gut

Conference: Microbiome Engineering

Year: 2020

Proceeding: 3rd International Conference on Microbiome Engineering (ICME 20)

Group: General Submissions

Session: Microbiome Therapeutics: Additive approaches

Authors:

Westereng, B., Norwegian University of Life Sciences
Pope, P., Norwegian University of Life Sciences
Michalak, L., Norwegian University of Life Sciences
La Rosa, S. L., Norwegian University of Life Sciences

Beneficial modulation of the gut microbiome has high-impact implications not only in humans, but also in livestock that sustain our current societal needs. In this context, we have tailored an acetylated galactoglucomannan (AcGGM) fibre to match unique enzymatic capabilities of Roseburia and Faecalibacterium species, both renowned butyrate-producing gut commensals. These bacteria have metabolic pathways dedicated to degradation of complex acetylated mannan. Here, we test the accuracy of AcGGM within the complex endogenous gut microbiome of pigs, wherein we resolve 355 metagenome-assembled genomes together with quantitative metaproteomes. In AcGGM-fed pigs, both target populations differentially express AcGGM-specific polysaccharide utilization loci, including novel, mannan-specific esterases that are critical to its deconstruction. However, AcGGM-inclusion also manifests a â€œbutterfly effectâ€□, whereby numerous metabolic changes and interdependent cross-feeding pathways occur in neighboring non-mannanolytic populations that produce short-chain fatty acids. Our findings show how intricate structural features and acetylation patterns of dietary fibre can be customized to specific bacterial populations, with potential to create greater modulatory effects at large.

Porous Microplates for Chemically Interactive Cell Culture

Conference: Microbiome Engineering

Year: 2020

Proceeding: 3rd International Conference on Microbiome Engineering (ICME 20)

Group: General Submissions

Session: Microbiome Therapeutics: Additive approaches

Authors:

Vaiana, C. A., Massachusetts Institute of Technology
Buie, C. R., Massachusetts Institute of Technology
Voigt, C. A., Massachusetts Institute of Technology
Kim, H., Massachusetts Institute of Technology

Microorganisms exchange information with each other and with their environments. Advanced co-culture techniques illuminate the importance of chemical diffusion between microorganisms. Existing techniques vary in complexity, degree of interaction, and experimental readout. We envision a culture platform that allows chemical exchange between wells while maintaining the simplicity of traditional multi-well plates. Herein we describe a microplate constructed of a porous co-polymer that allows multi-directional chemical exchange between spatially isolated culture wells. We use this microplate to culture a multi-member consortia of engineered *E. coli* and observe their response to six diffusing chemical inducers within the culture plate. We also demonstrate engineered quorum signaling and inhibitory peptide signaling between neighboring culture wells. These results imply the use of a porous microplate to study interacting liquid cell cultures. This platform can be applied toward designing engineered microbial consortia for distributed biosynthesis of therapeutics and other interesting compounds.

Xenosiderophore Utilization Promotes Bacteroides Thetaiotaomicron Resilience during Colitis

Conference: Microbiome Engineering

Year: 2020

Proceeding: 3rd International Conference on Microbiome Engineering (ICME 20)

Group: General Submissions

Session: Microbiome Therapeutics: Additive approaches

Authors:

Zhu, W., Vanderbilt University Medical Center

During short-lived perturbations, such as inflammation, the gut microbiota exhibits resilience and reverts to its original configuration. Although microbial access to the micronutrient iron is decreased during colitis, pathogens can scavenge iron by using siderophores. How commensal bacteria acquire iron during gut inflammation is incompletely understood. Curiously, the human commensal *Bacteroides thetaiotaomicron* does not produce siderophores but grows under iron-limiting conditions using enterobacterial siderophores. Using RNA-seq, we identify *B. thetaiotaomicron*genes that were upregulated during *Salmonella*-induced gut inflammation and were predicted to be involved in iron uptake. Mutants in the *xusABC* locus (BT2063-2065) were defective for xenosiderophore-mediated iron uptake *in vitro*. In the normal mouse gut, the XusABC system was dispensable, while a *xusA* mutant colonized poorly during colitis. Notably, *B. thetaiotaomicron* and *Salmonella* differs in capability to utilize siderophores, suggesting that certain siderophores could be engineered to serve as "Trojan horse" to specifically target pathogenic bacteria, while sparing the commensals. This work identifies xenosiderophore utilization as a critical mechanism for *B. thetaiotaomicron* to sustain colonization during inflammation and suggests a mechanism of how interphylum iron metabolism contributes to gut microbiota resilience.

Persistent Physiological Change in a Host Using Engineered Native Bacteria

Conference: Microbiome Engineering

Year: 2020

Proceeding: 3rd International Conference on Microbiome Engineering (ICME 20)

Group: General Submissions

Session: Microbiome Therapeutics: Additive approaches

Authors:

Russell, B. J., University of California, San Diego
Zarrinpar, A., University of California, San Diego
Brown, S. D., University of California, San Diego
Mai, I., University of California, San Diego
Dantas Machado, A. C., University of California, San Diego
Richter, R. A., University of California, San Diego
Pinto, A. F. M., The Salk Institute for Biological Studies
Miyamoto, Y., University of California, San Diego
Ho, S. B., University of California, San Diego
Eckmann, L., University of California, San Diego
Hasty, J., University of California at San Diego
Saghatelian, A., The Salk Institute for Biological Studies
Knight, R., University of California, San Diego

Engineered bacterial therapeutics have the potential to reverse disease by engrafting the gut lumen and providing a persistent beneficial effect on host physiology. However, attempts to functionally manipulate the gut microbiome of conventionally raised, wild-type (CR-WT) hosts have been largely unsuccessful. Here we show that specific and lasting functional changes in the gut microbiome can be achieved by isolating, modifying, and reintroducing native bacteria, thus allowing investigators to â€œknock inâ€□ individual functions into the gut microbiome. To demonstrate this approach, we isolated a native strain of *Escherichia coli* from stool cultures of CR-WT mice and modified it to express bile salt hydrolase (BSH), a bacterial enzyme that is theorized to affect host metabolism. Reintroduction of this strain by a single gavage induces perpetual, stable colonization of the entire gut, delivers the intended function, and affects host physiology. Furthermore, we demonstrate that tractable, native bacterial strains are easy to identify in humans and can be transformed to express a specific function. Thus, this approach to â€œknocking inâ€□ specific bacterial functions with engineered native bacteria allows for improved mechanistic studies of the role of the gut microbiome and makes individualized microbiome-mediated therapies with curative intent possible.

Unintentional and Intentional Engineered Probiotics Using Xanthan Gum

Conference: Microbiome Engineering

Year: 2020

Proceeding: 3rd International Conference on Microbiome Engineering (ICME 20)

Group: General Submissions

Session: Microbiome Therapeutics: Additive approaches

Authors:

Ostrowski, M., University of Michigan
La Rosa, S. L., Norwegian University of Life Sciences
Kunath, B. J., University of Luxembourg
Pereira, G., University of Michigan
Robertson, A., University of Michigan
Yao, T., Purdue University
Flint, G., University of Michigan
Pudlo, N. A., University of Michigan
Schmidt, T. M., University of Michigan
Hamaker, B. R., Purdue University
Pope, P., Norwegian University of Life Sciences
Martens, E., University of Michigan

Xanthan gum is a branched polysaccharide with complex sugar linkage structure and unique rheological properties making it useful as a food additive in a wide variety of processed foods at typical concentrations of 0.05-0.5% (w/w). Here, we show that the ability to utilize xanthan gum is surprisingly common in the individual human gut microbiomes of healthy adults and appears to be contingent on the activity of a single uncultured bacterium that is a member of an uncultured *Ruminococcaceae* genus. Using a metatranscriptomic approach followed by recombinant enzyme studies, we show that this keystone xanthan gum degrader depolymerizes xanthan gum with a novel glycoside hydrolase family 5 (GH5) enzyme that produces oligosaccharides that it can utilize via a series of additional enzymes. In some cases, oligosaccharide products of this pathway also cross-feed *Bacteroides intestinalis* strains equipped with their own distinct enzymatic pathway. Both xanthan gum loci were found with variable prevalence in industrialized human microbiome metagenomes but were absent in hunter-gatherers. Thus, consumption of a generally regarded as safe food additive, has apparently reinforced the existence of a food web between members of two different phyla of gut bacteria. Our findings have implications for the adaptability of the human gut microbiome to the introduction of new food additives on comparatively short timescales as well as utilizing the results of these events to engineer new orthogonal niches in members of the human gut microbiota for a future generation of probiotics.

Heterogolous FMT Facilitates Post-Antibiotics Microbiota Restoration but Does Not Guarantee Long-Term Maintenance in Aged Mice

Conference: Microbiome Engineering

Year: 2020

Proceeding: 3rd International Conference on Microbiome Engineering (ICME 20)

Group: General Submissions

Session: Microbiome Therapeutics: Modulatory approaches

Authors:

Dai, L., Shenzhen Institutes of Advanced Technology, Chinese Academy of Sciences

Fecal microbiota transplantation (FMT), a procedure to transfer fecal material from a donor to a recipient, has been shown to be an effective method to restore a healthy gut microbiota. However, the choice of donors may affect the therapeutic effects. Little is known about how the matching between donors and recipients influence the efficacy and long-term maintenance after FMT. In this study, we treated aged mice (20-month-old) with antibiotics and demonstrated that the gut microbiota diversity decreased significantly and did not recover to the original state. We carried out heterologous FMT for aged mice following antibiotics treatment from donors of different ages and found that heterologous FMT facilitated the restoration of microbiota diversity. For the recipient mice, we found that their long-term microbiota composition and gene expression profiles in colon were affected by the FMT donor. Furthermore, we tested whether the microbiota after FMT could remain stable under perturbations of colonic inflammation. We found that although recipient microbiota closely resembled the donor composition after FMT, the matching between donors and recipients is important for the long-term maintenance. Our study revealed the long-term effects of FMT on the microbiota of aged mice, suggesting that the matching between donors and recipients should be taken into account when FMT were implemented.

A Predictive Index for Health Status Using Species-Level Gut Microbiome Profiling

Conference: Microbiome Engineering

Year: 2020

Proceeding: 3rd International Conference on Microbiome Engineering (ICME 20)

Group: General Submissions

Session: Microbiome Therapeutics: Modulatory approaches

Authors:

Sung, J., Mayo Clinic

Providing insight into oneâ€™s health status from a gut microbiome sample is an important clinical goal in current human microbiome research. Herein, we introduce the Gut Microbiome Health Index (GMHI), a biologically-interpretable mathematical formula for predicting the likelihood of disease independent of the clinical diagnosis. GMHI is formulated upon 50 microbial species associated with healthy gut ecosystems. These species are identified through a multi-study, integrative analysis on 4347 human stool metagenomes from 34 published studies across healthy and 12 different nonhealthy conditions, i.e., disease or abnormal bodyweight. When demonstrated on our population-scale meta-dataset, GMHI is the most robust and consistent predictor of disease presence (or absence) compared to Î±-diversity indices. Validation on 679 samples from 9 additional studies results in a balanced accuracy of 73.7% in distinguishing healthy from non-healthy groups. Our findings suggest that gut taxonomic signatures can predict health status, and highlight how data sharing efforts can provide broadly applicable discoveries.

Gut Microbiota Composition Assessment Via Targeted Radical Polymerization (TRAP)

Conference: Microbiome Engineering

Year: 2020

Proceeding: 3rd International Conference on Microbiome Engineering (ICME 20)

Group: General Submissions

Session: Microbiome Therapeutics: Modulatory approaches

Authors:

Srivastava, S. K., Denmark Technical University

Drug delivery platforms that target the deposition of therapeutics to sites of disease are needed to increase efficacy and decrease off-target effects of current interventions. To address these issues, we developed a probiotic-based strategy capable of directly secreting therapeutic proteins into the gut lumen. Our chosen chassis *E. coli* Nissle 1917 (EcN) is a human probiotic with a proven safety record in human subjects. To enable EcN to secrete heterologous proteins into the gut lumen, we engineered it to encode a type III secretion system modified to secrete proteins into its surroundings rather than into host cells. Next, we developed variants of this strain, referred to as T3EcN, that secrete functional nanobodies, variable domains of heavy chain only antibodies (aka VHH), isolated from immunized alpacas and identified to sequester and neutralize target proteins of interest. This was done by appending the N-termini of the nanobody with a type III secretion sequence. To enable T3EcN to stably colonize and constitutively secrete nanobodies within the mammalian gut, nanobody-coding genes were placed under the control of constitutive synthetic promoters and introduced onto a plasmid that can be maintained in the absence of antibiotic selection. Notably, we observe that nanobody-secreting T3EcN variants stably colonize and maintain secretory activity within the mouse gastrointestinal tract. Furthermore, using a bioluminescent reporter for T3SS activity, we have demonstrated that T3EcNâ€™s T3SS is active in the colon and small intestine. Ongoing work in preclinical murine models is focused on establishing proof-of-concept for this novel targeted drug delivery platform.

The Death Phase in Microbiome Dynamics

Conference: Microbiome Engineering

Year: 2020

Proceeding: 3rd International Conference on Microbiome Engineering (ICME 20)

Group: General Submissions

Session: Microbiome Therapeutics: Modulatory approaches

Authors:

Verma, M.

Two approaches exist for developing microbiome therapeutics: *top-down* (which typically involves sequencing microbiomes from different physiological or pathological states and determining correlations between the microbial composition and their state) and *bottom-up* (which typically involves determining the role of individual microbes and interactions between these microbes for establishing causal relationships). The amount of data produced from sequencing (top-down) has increased tremendously in recent years. However, the inferences that researchers can make from these datasets have been largely limited to correlationsâ€"determining causation remains challenging or impossible. As such, microbiome research is unable to determine whether shifts in microbiome caused a pathological state or if a pathological state caused a shift in the microbiome. In contrast, because of the control they offer, bottom-up approaches have the potential to establish clearer insights about causation. Recent advances in culturing techniques and measurement have led to improved prediction of the behavior of complex communities. However, most model systems do not account for the death of microbes; since death of certain microbes could serve as nutrients for other microbes, it is an important component of microbiome dynamics. Here, I will share the effect of different monosaccharides and disaccharides on the growth and death of *Bacteroides fragilis* and *Escherichia coli* as two model organisms in a chemically defined medium. Quantifying the growth and death behavior has the potential to enable better modulation of microbiome therapeutics.

Engineering Multiplexed Inflammation Sensing Circuit for Gut Modulation

Conference: Microbiome Engineering

Year: 2020

Proceeding: 3rd International Conference on Microbiome Engineering (ICME 20)

Group: General Submissions

Session: Microbiome Therapeutics: Modulatory approaches

Authors:

Merk, L., Caltech
Shur, A., California Institute of Technology
Green, L. Jr., California Institute of technology
Murray, R. M., California Institute of Technology

The gut contains trillions of microbes that interact with host cells and monitor changes in the environment. Opportunistic pathogens exploit environmental conditions to stimulate their growth and virulence, leading to relapsing chronic disorders such as inflammatory bowel disease (IBD) in the gut. Despite the severity of these disorders, current therapies are effective in less than 30% of patients due to the lack of adherence to prescription schedules and broad, off-target effects. Fortunately, synthetic biology offers exciting opportunities to engineer smart microbial therapeutics that may colonize the gut, providing in situ surveillance and disease modulation. However, many current engineered microbes have been engineered to respond only to single gut environmental factors. In probiotic E. coli strain Nissle 1917, we implement the previously characterized split activator AND logic gate. Our system is capable of multiplexing two input signals, selective activation to an inflammatory biomarker, tetrathionate, and secondary input signal IPTG. We report 4-6 fold induction with minimal leak when both activators are present. We model the dynamics of the AND gate using chemical reaction networks, and by tuning parameters in silico, we identified perturbations that affect selectivity of our circuit. We anticipate that our results will prove useful for designing living therapeutics for spatial targeting and signal processing in complex environments.

Development of Designer Probiotics for Targeted Delivery of Immunomodulatory Payloads

Conference: Microbiome Engineering

Year: 2020

Proceeding: 3rd International Conference on Microbiome Engineering (ICME 20)

Group: General Submissions

Session: Microbiome Therapeutics: Modulatory approaches

Authors:

Lynch, J., Massachusetts General Hospital
Lesser, C., Massachusetts General Hospital

Drug delivery platforms that target the deposition of therapeutics to sites of disease are needed to increase efficacy and decrease off-target effects of current interventions. To address these issues, we developed a probiotic-based strategy capable of directly secreting therapeutic proteins into the gut lumen. Our chosen chassis *E. coli* Nissle 1917 (EcN) is a human probiotic with a proven safety record in human subjects. To enable EcN to secrete heterologous proteins into the gut lumen, we engineered it to encode a type III secretion system modified to secrete proteins into its surroundings rather than into host cells. Next, we developed variants of this strain, referred to as T3EcN, that secrete functional nanobodies, variable domains of heavy chain only antibodies (aka VHH), isolated from immunized alpacas and identified to sequester and neutralize target proteins of interest. This was done by appending the N-termini of the nanobody with a type III secretion sequence. To enable T3EcN to stably colonize and constitutively secrete nanobodies within the mammalian gut, nanobody-coding genes were placed under the control of constitutive synthetic promoters and introduced onto a plasmid that can be maintained in the absence of antibiotic selection. Notably, we observe that nanobody-secreting T3EcN variants stably colonize and maintain secretory activity within the mouse gastrointestinal tract. Furthermore, using a bioluminescent reporter for T3SS activity, we have demonstrated that T3EcNâ€™s T3SS is active in the colon and small intestine. Ongoing work in preclinical murine models is focused on establishing proof-of-concept for this novel targeted drug delivery platform.

Bacterial Resistance to CRISPR-Cas Antimicrobials

Conference: Microbiome Engineering

Year: 2020

Proceeding: 3rd International Conference on Microbiome Engineering (ICME 20)

Group: General Submissions

Session: Microbiome Therapeutics: Subtractive approaches

Authors:

Vazquez Uribe, R., DTU
Sommer, M., Technical University of Denmark

In the age of antibiotic resistance and precise microbiome engineering, CRISPR-Cas based antimicrobials promise to have a substantial impact on the way we treat diseases in the future. However, the efficacy of these antimicrobials and their mechanisms of resistance remain to be elucidated. Here, we systematically investigated mutations that arise from CRISPR-Cas9 induced death. We designed a genetic circuit using Cas9 from Streptococcus pyogenes (SpCas9) to study the different parameters that affect the efficiency of the system and elucidated the spontaneous genetic alterations that confer resistance to this novel type of antimicrobial agent. Our findings highlight that killing efficiency is not correlated with the type of target or number of cutting sites, however we observed that multiple target sites are valuable in order to prevent genomic mutations that can result in resistance. Furthermore, we observed that in most cases SpCas9 is the preferred target for mutations inactivating the activity of the antimicrobial by large-scale bacterial genome rearrangements involving mobile genetic elements such as insertion sequence (IS) elements. Finally, we demonstrate this can be overcome by simulating a multi-dosing strategy that resensitize escaper cells. The work presented here provides a guide to design strategies that reduce resistance and improve the activity of CRISPR-Cas antimicrobials.

In silico Simulation to Predict Activity of a Synthetic Biotic, SYNB8802, in Healthy Volunteers and Patients with Enteric Hyperoxaluria

Conference: Microbiome Engineering

Year: 2020

Proceeding: 3rd International Conference on Microbiome Engineering (ICME 20)

Group: General Submissions

Session: Microbiome Therapeutics: Subtractive approaches

Authors:

Horvath, N.
Lubkowicz, D., Synlogic
James, M., Synlogic Inc.
Cantarella, P., Synlogic
Charbonneau, M., Synlogic Inc

Enteric Hyperoxaluria is an acquired metabolic disorder caused by increased absorption of dietary oxalate, which is present in many common foods including leafy greens, nuts, and chocolate. Enteric Hyperoxaluria often occurs as a result of a primary insult to the bowel, such as inflammatory bowel disease, Crohnâ€™s disease, short bowel syndrome, or as a result of surgical procedures such as Roux-en-Y bariatric weight-loss surgery. Enteric Hyperoxaluria results in dangerously high levels of urinary oxalate, which causes progressive kidney damage, kidney stone formation, and nephrocalcinosis. There are an estimated 250,000 patients with Enteric Hyperoxaluria in the United States and no approved treatment options. SYNB8802 is an engineered non-pathogenic strain of E. coli(Nissle), using Synlogicâ€™s Synthetic Biotic platform, designed to consume oxalate in the GI tract and lower urinary oxalate levels, potentially reducing kidney damage due to Enteric Hyperoxaluria. We present an approach for in silico simulation (ISS) of SYNB8802 activity based on in vitro gastrointestinal simulations and the relevant disease biology. This computational framework considers the dynamics of metabolite availability and microenvironmental conditions to predict the capacity of SYNB8802 for consumption of dietary oxalate and subsequently reduction of plasma absorption and urinary excretion of oxalate. For Enteric Hyperoxaluria patients consuming a reasonable dietary intake of oxalate, ISS predicts a dose-dependent lowering of urinary oxalate excretion at a clinically meaningful level. Along with in vitro gut simulations and animal models, the ISS approach contributes to a physiologically based understanding of SYNB8802â€™s therapeutic potential for Enteric Hyperoxaluria patients.

AUTHOR INDEX

Notes:

Notes:

Notes:

Notes:

AIChE
120 Wall Street, FL 23
New York, NY 10005-4020

ISBN 978-1-7138-2773-3

9 781713 827733

CURRAN ASSOCIATES INC.
proceedings.com